ABOVE: *Rivers were specially dammed for retting flax, or oblong holes, known as 'lint holes', were dug in boggy areas. In Northern Ireland retting was sometimes called 'bogging'.*

COVER: *An engraving of 1783, by William Hincks, showing finishing procedure in a bleach mill. (See page 28.)*

FLAX AND LINEN

Patricia Baines

Shire Publica

CONTENTS

Set in 9 point Times roman and printed in Great Britain by C. I. Thomas & Sons (Haverfordwest) Ltd, Press Buildings, Merlins Bridge, Haverfordwest, Dyfed.

British Library Cataloguing in Publication data available.

ACKNOWLEDGEMENTS

The author wishes to thank the following for granting permission to publish illustrations: International Linen Promotion, pages 5 (upper), 15 (lower); Museum für Deutsche Volkskunde, Berlin, pages 6 (lower), 7 (lower), 13 (upper right); Nationalmuseet, Copenhagen, pages 6 (upper), 12 (upper), 31; National Museum of Antiquities of Scotland, Edinburgh, pages 9 (lower), 22 (lower), 24; Science Museum, London, pages 9 (upper), 10 (upper), 16 (upper left); Shirley Institute, Manchester, page 4 (left); Ulster Folk and Transport Museum, County Down, pages 1, 3, 5 (lower), 7 (upper left), 15 (upper), 17, 27, 28 (lower) and cover, 29; Ulster Museum, Belfast, pages 7 (upper right), 8, 19, 28 (upper); Victoria and Albert Museum, by courtesy of the Board of Trustees, pages 18, 21, 23, 25, 26; Colin Wilson, page 20. The drawing on page 14 is by D. R. Darton.

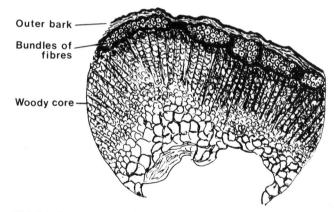

ABOVE: *A cross-section of a flax stem showing the little bundles of fibres lying just beneath the outer bark; there are between about fifteen and forty bundles (with a similar number of fibres per bundle) in each stem.*

Outer bark
Bundles of fibres
Woody core

LEFT: *The plant Linum usitatissimum has narrow pointed leaves and a five-petal blue or white flower. The wild form of flax, Linum angustifolium, is a common weed in the Mediterranean area and may have derived from the cultivated type. There are a number of wild varieties (including Linum angustifolium) to be found in Britain. Ornamental garden varieties of flax should not be confused with the crop variety and are not suitable for processing.*

Flax is always pulled up (never cut) to preserve the maximum length of the fibres and also to prevent damage during the subsequent process. When done by hand it was a back-breaking job.

FLAX

For many thousands of years flax has been a valuable cultivated crop for two purposes, the seed and the fibre. Since early man was a seed gatherer, it seems likely that the flax plant first attracted his attention for the seed: not only is it edible but it can be pressed to produce oil (linseed). Since it is thought that weaving preceded spinning, the early seed gatherers may have recognised the toughness and flexibility of the flax stalks and collected them too, to be used for intertwining with stakes driven into the ground to make wind breaks, or across streams to make fish traps. Thus, after continual exposure to the elements, alternately wet and dry, the outer bark of the stalks would have rotted, revealing the shiny fibres and attracting man's attention.

Flax is one of the dicotyledonous plants (those that form two seed-leaves, not one) that produce bast fibres (that is, coming from the stem), which run from the root of the plant to the tip. The particular characteristics of flax fibres give linen its special quality. They are long, strong, lustrous, readily absorb water, are a good conductor of heat but lack elasticity. Flax is an adaptable plant and, given the rich loamy soil that it needs, can tolerate very different climates in many parts of the world. Northern France, Belgium and Holland have, for centuries, been famous for fine quality flax and still continue to grow it, but by far the greatest output today is from the USSR. China grows flax, as do the Baltic countries, including Poland, which have a long history as producers of flax and seed, while Egypt, so famous for flax in past millennia, is now one of the smallest contributors to the world market. Other countries such as Sweden, Northern Ireland and Scotland, once well known for flax, have gradually ceased to grow it as a commercial crop, although experiments in Scotland and Ireland are in progress and an increasing number of farmers in Scotland are now growing flax for a new mill in Arbroath.

When flax is grown for fibre the seeds are sown close together so that the stems grow straight, with as few branches as

3

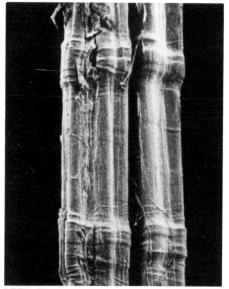

LEFT: *The fibres are composed mainly of cellulose with short overlapping cells held together with a gummy matter. Under the microscope the individual fibre is seen to be long, smooth and cylindrical but with slight bumps on the surface where the cells overlap.*

RIGHT: *Hemp, Cannabis sativa, is an annual of the Cannabaciae family (the subspecies indica yields the narcotic). Like flax, hemp is a bast fibre growing, however, to a height of at least 10 feet (3 m). It spread from its original habitat in central Asia and adapted itself to a variety of rich soils in many parts of the world, Italy producing some of the finest qualities. Hemp is processed in the same way as flax but requires harder treatment since the fibres are coarser.*

possible, to achieve the maximum fibre length. Sowing is done usually in the spring, after the winter frosts. The growing period is about three months, by which time the plants are 3 to 4 feet (910-1220 mm) high and the lower parts of the stems are beginning to turn yellow. To obtain the finest quality fibre, the seeds should be a little under-ripe when the flax is harvested. It was often the practice to leave a small patch of the crop to ripen fully, to provide seed for the next year.

When the crop has been pulled up the stalks are gathered in bundles *(beets),* tied up and either stacked *(stooked)* or hung up to dry for about two weeks.

To separate the fibres from the rest of the plant, it is necessary to decompose the woody matter and cellular tissue surrounding the fibres. There are several methods: one is to lay the flax in thin layers flat on the grass in rows, known as *dew-retting,* which takes from twenty to thirty days. If the weather is dry, there is insufficient dew for the necessary ret, so the farmer has to water the flax to keep it moist. It needs experience to judge exactly when the right amount of retting has taken place before causing damage to the fibres. Dew-retted flax is generally darker in colour than water-retted. In *water-retting* the bundles of flax are submerged in pools or gently flowing soft-water streams for about ten to fourteen days. Retting, however, pollutes the water and creates a very unpleasant smell. More modern methods utilise specially built tanks and carefully controlled temperatures. Fine quality flax may be double-retted by first putting it partly through the process, drying it and retting it again. Chemicals are also used and the latest method is to spray the crop with a special herbicide before it is pulled, although much of the Belgian crop is dew-retted since this is the least costly method.

Once the correct amount of decomposition has taken place, the flax is completely dried. It can then be stored, ready for the series of processes known as *flax dressing.*

4

ABOVE: *Much of the flax harvesting machinery was designed in Belgium. Motorised pulling machines, such as the one shown here, have dividers in front to guide the plants to the pulling unit (a rubber belt and a drum), which grips the stems and draws the plants vertically from the ground. When the flax reaches the top of the drum, another set of belts deposits it on the ground on the off side of the machine.*

BELOW: *The seeds were removed by passing the heads of the stalks through a coarse comb, the ripple. A cloth under the bench gathered the seeds for linseed oil production or cattle cake. Machines are now used for de-seeding in flax processing factories; before their invention, when labour costs made hand rippling too expensive, the seeds were left on during the next process of retting, losing much of the linseed crop.*

5

The flax breaker is a chopping device with two or three blades, pivoted at one end on a stand with slots for the blades to fall into. Bundles of flax, held at the root ends and constantly turned, are pounded by the blades to break up the straw. Hand breakers were usually of two types of design: either operated by an extended handle or from on top of the blades (as illustrated). This less usual design has a bench extension for stacking the straw.

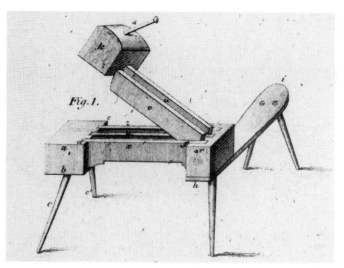

Fig. 1.

First, the straw has to be *broken*. The most primitive method is to hit the whole length of the stalk with a mallet, but a simple piece of equipment, the flax breaker, was invented in the fourteenth century, probably in Holland, and came to be more generally used when the

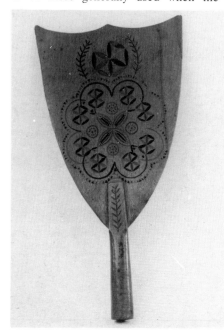

breaking was done by hand.

To remove the broken straw (known as *boon, shoves* or *shous*) the bundles are *scutched*. The hand method uses a vertical wooden board with a slot cut in the top or the side. The flax is beaten against the board with a scutching bat or *blade*, which removes the boon. Scutching is sometimes called *swingling*.

Lastly, the flax is *hackled* by passing the bundles of fibres through a series of combs to remove any small remaining pieces of straw, to remove the short fibres (*tow*), and to align the long fibres (*line*) parallel. Line, which is made into bundles known as *stricks*, is used to produce the finest yarn. Tow makes a rougher, coarser yarn.

A German 'Schwingelblatt'. In many parts of Europe tools used in flax dressing, as well as spinning wheels and distaffs, were given as wedding presents, often beautifully decorated or carved.

ABOVE, LEFT: *From the early eighteenth century machinery for breaking and scutching was, in some areas, driven by water. In the mid nineteenth century this type of breaker was developed by the McAdam brothers of Belfast and used in the Ulster mills. The machine consisted of five pairs of fluted metal rollers, each progressively finer in depth and width, placed side by side horizontally. The bundles or stricks of flax straw were fed by hand to the coarsest pair of rollers and emitted from the finest. The broken straw was then collected and re-stricked ready for the scutchers.*

ABOVE, RIGHT: *To replace the hand scutching blade a water-driven revolving circle of wooden blades (called 'handles', the number varying from four to twelve) hit the bundle of flax held in the scutcher's hand. Protected only by a wooden board, scutchers often lost their thumbs. There were over nine hundred small scutching mills in Ulster in the mid nineteenth century, working from four to twelve 'births'. Scutching created a great deal of dust and at this stage quite a lot of the fibre went for tow.*

RIGHT: *Hackle pins were usually clustered in square, oblong or round blocks, the two or three blocks graduated from coarse pins, fairly wide apart, to fine pins, closer together. They were inserted into a wooden block and mounted on a bench or, in Germany and Holland, on to boards placed in stands (as illustrated here).*

7

RIGHT: *Inside the York Street flax spinning mill in Belfast. The stricks of flax being given a preliminary hackle (known as 'roughing') with a coarse comb to straighten the fibres.*

BELOW: *In the hackling machines the stricks of flax are clasped so that they hang vertically. As they move horizontally along one side of the machine they pass revolving belts carrying the hackle pins, which pass through the fibres. Successive belts carry pins that are finer and closer together. At the far end of the machine each strick is turned over so that its other half is hackled as it returns along the opposite side of the machine. The tow falls into a trough below.*

8

SPINNING FLAX INTO LINEN

Spindle weights (*whorls*) have been found in excavated neolithic sites in Syria, Mesopotamia (Iraq) and Persia (Iran), with fragments of flax fibres, and so flax cultivation and spinning can be dated at least to 8000-6000 BC. There is also evidence that the neolithic lake dwellers in Switzerland spun flax and wove linen.

Egyptian tomb paintings dating back to about 1900 BC provide valuable information about flax spinning methods. Spinning is a threefold operation: drawing out (*drafting*) the fibres, twisting them into a continuous yarn and winding on the yarn to stop it untwisting. The Egyptians were highly organised, a single task being allotted to each person, so that the workers were little more than human machines. The drafting of the flax fibres was done by one person, sitting cross-legged on the ground, the fibres sliding past one another a few at a time and lying parallel before being given a light twist between the palm of the hand and the thigh. A second person twisted the pre-pared fibres (*rovings*) into a strong yarn, using a spindle with a whorl at the top of the shaft and a hook above to hold the yarn. The yarn was wound on round the shaft just below the spindle whorl.

Spindles used for flax spinning in Europe generally had the whorl at the bottom end of the shaft and a distaff was

LEFT: *A tomb painting at Beni Hasan depicts dextrous ancient Egyptian spinners who could use two spindles simultaneously. The prepared ribbons of fibres were rolled into balls and placed in a bucket behind the spinner. The spindle was twirled by rolling the shaft along the hip; the rotating spindle was then tossed into the air, suspended by the yarn. By standing on a platform, long lengths of yarn were spun before stopping to wind on.*

RIGHT: *A young Scottish girl spinning flax with a spindle. The use of the distaff tucked under the spinner's arm (or some-times into a belt round the waist), enabled the spinner to pick up the work whenever there was a spare moment and spin as she moved about. Distaffs vary from country to coun-try, but for flax they often took the form of a fairly long pole with some sort of cone or 'cage' at the top.*

9

used to support the prepared fibres. To make a fine smooth thread, the flax fibres were wetted to loosen the gummy matter. Saliva was most convenient for doing this and in some traditions the linen yarn was passed through the mouth. In others the thumb was licked before smoothing the thread.

The earliest type of spinning wheel, which came from the east, was unsuitable for spinning flax, but the flyer spinning wheel, which probably was not introduced until the second half of the fifteenth century, was intended for spinning flax. This important development (which appears to have taken place in Europe) depends on the difference of speeds in the rotation of the bobbin (loosely fitted on the horizontal spindle) and the spindle itself, making twisting and winding on simultaneous and spinning therefore continuous. Flyer spinning wheels are often referred to as flax wheels (in the early days, linen wheels), long fibre wheels or Saxony wheels (though there is no evidence that Saxony was their place of origin). The early wheels were turned by the right hand while the left hand drafted the fibres from the prepared distaff, either attached to the spinning wheel itself or free-standing. There are many shapes and types of distaff and the method of dressing them varies. The fibres are often criss-crossed over each other before being tied to the distaff; alternatively they can hang straight down or be simply wound round the distaff.

A seventeenth-century flax spinner from Holland. Since she is using two hands to draft the fibres, a treadle can be assumed. Dutch spinning wheels were introduced into Ireland by the Earl of Strafford in 1632. At the very end of the century Louis Crommelin was appointed overseer of the Irish linen industry. He deplored the Dutch wheels because the women would not stop treadling when there was a lump in their thread.

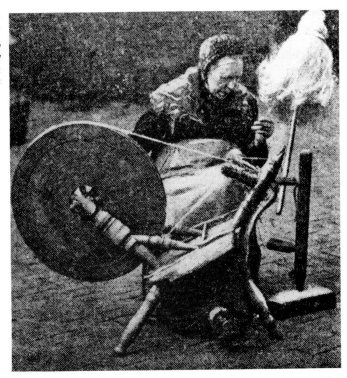

A late nineteenth century spinner from Courtrai, Flanders, using a spinning wheel with a disc wheel turned by hand. Much of the finely spun yarn came from this area.

During the early seventeenth century a treadle was introduced, linked to the axle of the spinning wheel, so that the spinner had both hands free for drafting the fibres. (Even so, in many parts of Europe the tradition for spinning flax with the left hand only persisted.) The introduction of the treadle also made it practicable for the flyer mechanism to be placed above the wheel, making a more compact tool.

Both vertical and horizontal spinning wheels were used for flax. Often the wheels were fairly small (14 to 18 inches; 360 to 460 mm). There are other features which identify flax wheels, besides a distaff or an arm or hole for holding a distaff. One such feature is a recess to put a water pot (or a little cup attached to the stock) for wetting the fingers, a more healthy and lady-like method of wetting flax than the use of saliva. Other signs are plenty of small bent wires on the flyer arms and deep grooves cut into the wood by the hard yarn.

Two-handed spinning wheels used for flax are known from the last quarter of the seventeenth century. Such wheels were introduced to help the poor to increase their earnings, often being worked by little girls of six to seven years old. Charity schools were set up to teach children to spin flax, particularly in areas where the linen industry was being encouraged. In Scotland two-handed spinning wheels seem to have been used quite widely and with success, but this was not always the case.

The *reel* was an important piece of equipment and often sold as a pair with a flax spinning wheel. To avoid having the spinning wheel idle, another member of the family reeled off the full bobbin into a *hank*. The hanks were subdivided into *leas* and the number of leas spun from 1 pound (0.45 kg) of flax gave the count number of the yarn, the greater the number of leas the higher the count number and the finer the yarn. Count tables varied from place to place, being

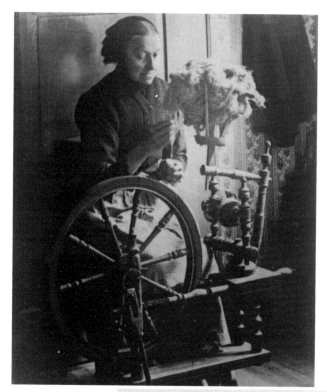

LEFT: *Spinning from a tow distaff in Denmark. A bunch of tow is placed on the prongs of the distaff and when spun makes a rougher type of yarn. At least a third of the flax after scutching and hackling is tow. With mechanisation tow is carded: a special type of carding machine suitable for tow was developed in 1820 by William Brown of Dundee.*

RIGHT: *In the 1780s, after the introduction of spinning machines in the cotton industry, there arose a fashion to spin flax. Ladylike 'boudoir' spinning wheels from this period were highly decorative. The one illustrated here is a table model without treadle which is balanced across the knees of the spinner. A large apron protects her gown from becoming covered with the fine dust that falls from the flax.*

ABOVE, LEFT: *This earliest known picture of a two-handed spinning wheel was published at the beginning of a letter written by Thomas Firmin in 1681. Firmin's business produced coarse canvas pepper bags. He was a philanthropist who ran a workhouse for children to spin the yarn using this type of spinning wheel. Firmin lost about £200 a year by this enterprise.*

ABOVE, RIGHT: *Click or clock reels were fitted with a simple mechanism to indicate that the reel had turned a certain number of times and, from the circumference, the yardage could be calculated. After each click the group of threads (lea) was tied together and another group reeled off. One arm of the reel was removable or collapsible for easy removal of the hank.*

LEFT: *To prepare the fibres for the spinning machines the bundles of hackled flax are placed on a spread board, a continuous moving leather belt, which feeds the fibres to the first drawing rollers, producing a thick continuous ribbon of fibres. The ribbons are then doubled and drafted repeatedly until a fine uniform roving is ready for the spinning machines.*

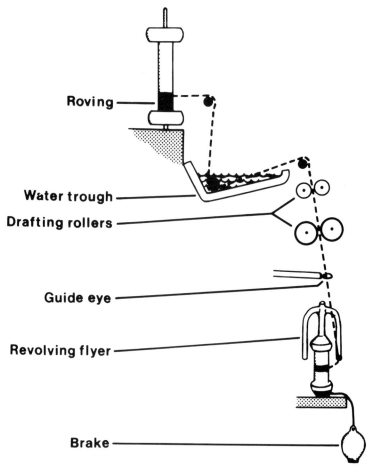

Roving

Water trough

Drafting rollers

Guide eye

Revolving flyer

Brake

---- Path of roving into yarn

The position of the water trough in wet spinning.

based on different reel circumferences, and this could be very confusing for manufacturers. Today 300 yards (274 m) is the standard lea for linen.

Early attempts to mechanise flax spinning were not successful, the long fibres making it difficult. The first firm to achieve success was Marshall and Company of Leeds, whose foreman mechanic, Matthew Murray, patented machinery in 1790, which, however, could spin only the coarser counts. For this reason there was work for hand spinners who could spin the fine counts well into the

nineteenth century. Much of the fine yarn was imported from Holland and Germany.

In 1814 a Frenchman, Philippe de Girard, patented a wet spinning machine which used three rollers half submerged in a trough of water. Significant improvement came in 1825 when James Kay of Preston discovered that if water in the trough was hot it softened the gummy matter in the flax quickly, making it possible to draft the fibres into fine count yarns.

ABOVE: *A wet flyer spinning frame of the early twentieth century; the water trough is heated almost to boiling point.*
BELOW: *Cops of line yarn ready to go forward to the reeling room where they will be wound into hanks for further processing.*

BELOW: *The interlacing of threads for plain weave.*

ABOVE, LEFT: *Horizontal ground loom, as depicted in a tomb painting from Beni Hasan, Egypt, about 1900 BC. The warp is stretched taut between two wooden beams, each supported by two short stakes driven into the ground. To make the plain weave cloth, a thin wide stick (shed stick) is placed under every alternate warp thread. Turned on its side, the stick creates an opening (the shed) through which the weft yarn is passed from one side to the other. To form the counter shed, alternate warp threads are looped (with cord) on to a second stick (heddle rod) which lies across the surface of the warp. When the rod is pulled up it raises the warp threads tied to it, and the weft is again passed through the shed. After each interlacement, the weft is pushed (or beaten) along the warp with a wooden sword (by the two women crouched on the ground) to form cloth at the near end of the loom.*

BELOW: *The treadle loom appears to have been introduced around 1000 AD; though not specially for linen, it was eminently suitable. An advantage of this type of loom is that, by means of pulleys and cords, the shafts carrying the warp threads are attached underneath the loom to the treadles, which, when depressed, raise first one shaft carrying alternate threads while the other treadle raises the second shaft. This leaves the weaver with both hands free for throwing the shuttle (carrying the weft) backwards and forwards through the sheds. Another advantage for linen production is the counterbalance method of shedding. When one set of threads is raised the other set is pulled down so there is an even stress on the warp thread at every change of the shed. Whereas wool, with its elastic properties, will spring back if extra strain is put on a particular group of threads, linen will not.*

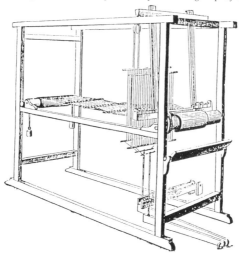

With the treadle loom the warp passes through little eyes in string heddles (slung on the shafts), which may rub the threads. If the warp is fine, and particularly if it is linen, this causes the threads to fray and break. To overcome this difficulty, a dressing or 'size' is applied to the warp. This is a thin glutinous solution, often made from boiling flour and water, which is brushed on to the part of the warp that lies between the back beam and the heddles. This slowed the work of the weaver since he was constantly stopping to size the warp and wait for it to dry. It was not until the beginning of the nineteenth century that a machine was invented which applied the dressing after the warping process, so that it was dry before reaching the loom.

WEAVING LINEN CLOTH

Weaving cloth on a loom can be dated to at least the fourth millennium BC through the discovery during excavations in Lower Egypt of a pottery bowl with a representation of a loom painted on it. It can be assumed that it was the type used for linen since a similar loom was found painted on the wall of an Egyptian tomb over two thousand years later. The regulated teamwork of the Egyptians in making linen cloth resulted in large quantities being produced, its fine quality continuing to be admired and prized for many hundreds of years. The cloth was predominantly plain weave but, since there was no method of keeping the warp threads (those running lengthwise) evenly spaced across the width of the early type of loom, much of the cloth was warp rep, that is, with more warp threads showing than weft threads (those running widthwise). The finely woven shrouds of the Egyptians and their mummy wrappings, linen torn into strips which bound the embalmed body to protect it against destruction by insects, are well known. Thicker linen was used for stuffing and reshaping the body after removal of entrails. Linen is an ideal fabric for hot climates; its readiness to absorb water, and therefore sweat, makes it a hygienic material to wear next to the skin and, because it is a good conductor of heat, it has a cool feel. Linen creases easily owing to the non-elasticity of the fibre, a tendency which the Egyptians exploited

A shirt of the sixteenth century, in the Victoria and Albert Museum, made of very fine, smooth linen and embroidered at the cuffs and neck with silk.

by deliberately creasing the cloth while it was wet to make concertina-like pleats. This gave it 'spring', making very finely woven cloth cling to the body and so appear transparent. Pleated tunics made of such cloth were worn by rich women and dancing girls while heavier pleated kilts were worn by men of rank. Every time the garment was washed it needed to be re-pleated.

More closely woven linen cloth was worn by the Phoenicians as armour, since its toughness acted as protection from the dangers of war and hunting yet gave the wearer greater mobility than when he was dressed in chain mail.

In Egypt it was believed that flax was created by the goddess Isis and the priests who served her were obliged to wear nothing but clean white linen as a symbol of divine purity. The importance of linen for religious rights continued with the Israelites, and the custom was maintained by both the Greeks and the Romans. The liturgical vestments of the Catholic church included a long white tunic (alb) and an oblong piece of white linen cloth (amice) worn, originally, over the head and neck and often taken to symbolise 'the helmet of salvation'.

With the expansion of the Roman Empire linen production reached many parts of Europe. Pliny tells of the fineness of the linen made in Tarragona (Spain) because of the excellent washing water there. He also praises the linen made in Italy and describes the awnings made of linen that were stretched over the theatre of the games of Apollo and over the Roman forum. In Germany, he says, linen made the most beautiful dress material and the women there made the cloth in 'caves dug underground'. This underlines the necessity for humidity when weaving linen.

Throughout Europe from the early middle ages to the end of the eighteenth century the cultivation of flax and the making of linen would have been widespread domestic occupations, each household growing sufficient flax for its own needs. Often tithes and rents were paid with bolts of linen cloth. The processing and spinning was done by the women and sometimes also the weaving, but it was equally likely that a journeyman weaver would pay an annual visit to make the cloth from the family spinning. Such traditions could be found in many parts of Europe well into the nineteenth century, in some isolated areas even into the twentieth century, and were also reflected in the pattern of life of settlers in North America. One of the most deep-rooted of these traditions was the making of a chest full of linen for a bride's dowry, for which only the best the family could produce was good enough. Hand-spun, hand-woven sheets often had the initials

18

Weaving linen on a power loom proved difficult because of the inflexibility of the fibre; this meant that linen production lagged behind the cotton industry, and not until the beginning of the nineteenth century were factories set up to weave linen. With the shuttle thrown backwards and forwards automatically, it was possible to employ young girls to tend looms. Their job was to piece (i.e. mend) broken warp threads and keep the shuttles supplied with spools of weft yarn. A man known as a 'tenter' was the skilled mechanic who was responsible for the maintenance of a group of looms.

of the bride and groom, and sometimes the date, embroidered along one edge.

The shirt, shift, sark, chemise or smock were names given to body garments worn next to the skin and for this reason they were usually made of linen. They were worn by both sexes, shifts and chemises usually being women's clothes and shirts being for men (not until the nineteenth century was an undershirt or vest worn). Shirts for the well-off were made of very fine smooth linen while poorer people wore those made of hemp and tow yarn, which were rougher, coarser and probably not very comfortable. The neckline of the chemise or shirt, drawn in by a string to form a small ruffle or frill, was the precursor of the ruff with its extravagant use of fine plain linen. Linen was used extensively, particularly by women, for covering the head through successive periods of fashion. Kerchiefs worn round the shoulders and aprons were usually made of linen while much coarser types were used for linings, interlinings and for stiffening skirts. The fashion for fine linen handkerchiefs, often embroidered with silk and later trimmed with lace, started in France in the sixteenth century. The early bobbin lace (deriving from weaving) was made with very fine linen thread (plied linen yarn) and linen cloths of different weights have been, and still are, used as a base for all types of embroidery.

Famous for linen weaving were Flanders, Holland and northern France. *Toile de Reims* was reputed to be as fine as Egyptian linen and from the same area came other very fine plain weave linen; *cambrics,* named after Cambrai, and *lawns* named, perhaps, after the neighbouring town of Laon.

Linen weavers from the European mainland, particularly the Low Countries and France, settled in various parts of the British Isles, fleeing from religious persecution. Sometimes they brought with them new equipment and techniques which stimulated the trade. Usually they chose to settle where the linen industry was already well established but they often did not integrate with the local community. England's linen industry was overshadowed by the important wool industry, which was jealously protected. Because of this, in the seventeenth and eighteenth centuries the linen industry in Ireland and Scotland was especially encouraged. A Board of Trustees was set up in Dublin in 1711 and in Edinburgh in 1727. Both boards had funds from which bounties could be given to finance all aspects of the industry, including the provision of better seed for the farmers, teaching women to spin flax, and setting up bleach greens. They also established standards, with inspections of woven cloth and stamp masters. Such standards included uniform widths and lengths to protect buyers from dishonest weavers. As well as the fine linen that was made, medium and coarse linens accounted for much of the industry, for sheets, shirts and all manner of household needs. Millions of yards of canvas and duck were needed every year for sail cloth (for which the long flax — line — was preferred but not always used). Coarser cloths for, amongst other things, wagon covers, grain bags and sacks were made of tow or hemp (these were also used for poor people's clothes and sheets) and hemp was the best fibre for ropes and cordage. Painting on linen goes back to Egyptian times and linen canvas is still favoured by artists today.

In the west linen was the most widely used cloth made from a vegetable fibre for a long period until the mechanisation of cotton spinning in the last quarter of the eighteenth century. From the time of the industrial revolution there was a gradual decline in the demand for linen; cotton cloth, being cheaper to produce as well as lighter to wear and easier to launder, was being manufactured in vast quantities. There were unexpected booms for linen such as at the time of the American Civil War, when there was a shortage of cotton, and there were many uses for which cotton was not a substitute for linen, particularly when toughness and durability were required. Nevertheless the pre-eminence of flax as the most important vegetable fibre was lost.

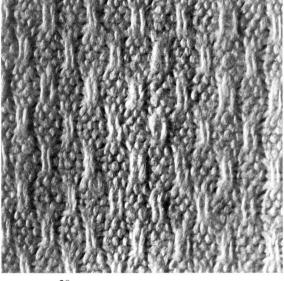

Detail of a hand towel from Norway woven in huck-a-back pattern (other names include 'huggabag', 'huckabag' or simply 'huck'). An old weave still in use today for hand towels, the uneven surface produced by the weave pattern absorbs water and dries off quickly. Many derivations of this weave are found in folk weaving, particularly from Germany and Scandinavia. The name may have been adapted from 'hucksters', known from around 1200, who were pedlars who carried their wares on their back and sold linens in the market places.

20

'Perugia towels' of the sixteenth century have blue and white figurative border patterns. Surviving until quite recently as a folk weave in Italy, these patterns are considered to be the precursors of the linen damasks, although the weaving technique is not the same.

PATTERN WEAVES AND DAMASK

When the four-shaft treadle loom became established in the late middle ages (the exact date being unknown) it was easy to weave small repeat patterns (controlled by the loom) as fast as plain weave. These patterns are based on what is known as *twill* weave. Its characteristic is a diagonal line appearing as part of the weave structure. There are an immense number of variations possible with twill weave, producing zigzag, chevrons and diamond patterns. From the fifteenth century onwards the word *diaper* is generally used to describe linens which are woven with all-over patterns based on the twill structure. They can vary from the very simple to complex patterns which require either a great number of shafts and treadles or a more complicated loom. Because of the smoothness and hardness of the yarn, patterns woven in linen show clearly and effectively.

Perhaps the most important products of linen weavers, for which their technical skills and imaginations were stretched, were table cloths and table napkins. In the middle ages white cloths were placed on altars, covered whole tables for important feasts or part of a table for important people. Early decorative work appeared first in the borders of table cloths, consisting of openwork and fringes, some sort of embroidery or woven blue bands. By the fourteenth century all-over diaper patterns began to appear.

In the sixteenth century table cloths with large complicated figurative patterns became fashionable in the royal courts and by the seventeenth century they were an expensive status symbol of the upper classes. Such cloths, often with napkins to match, were woven to commemorate historical events such as victories or coronations or to mark a state occasion or important marriage. Mythical and religious subjects and hunting scenes were favourite themes. Incorporated in the designs would be crests, insignia, dates, names of people and places, including sometimes the name of the place where a cloth had been woven. Decorative borders, often with floral or fruit motifs, sometimes stylised, surrounded the central subject. These cloths, made of linen, are called *damasks* and were woven on a *drawloom*.

There is much uncertainty as to when and where this complex loom first appeared, but it seems most likely to have been in the first century AD in

The interlacing of threads for twill weave.

The interlacing of threads for satin weave. The ground weave for damask requires five shafts and avoids the diagonal line of twill.

BELOW: *A Scottish drawloom weaver. The weave required two sets of harnesses, one behind the other, working in conjunction with one another. The front harness (nearest the weaver) produced the ground weave based on a five-shaft satin weave. The heddles had elongated eyes so that the warp threads (ends) could be raised by the back harness as required. The front shafts were worked by treadles. Each heddle (or mail) of the back harness was independently weighted. The loom required two people: the weaver sat in front of the shafts, throwing the shuttle and treadling the ground weave; a 'draw boy', by pulling cords, lifted the groups of heddles required for the pattern by the weaver. In the east, the draw boy sat above the pattern harness while in Europe he stood to the side.*

A detail from the centre of a Flemish linen damask cloth in the Victoria and Albert Museum dating from the end of the fifteenth century, depicting the sacrifice of Isaac. The earliest linen damask designs imitated those used in the silk industry but biblical scenes (religion being an all-pervading influence) were popular in the sixteenth century, the patterns apparently being derived from wood cuts.

Persia, though some scholars believe it to have been a little earlier and in China. It was for weaving silk that damask (named after the city of Damascus in Syria) is famous and for which it was developed.

Exactly when linen damask first made its appearance is not known but as a result of the crusades Europeans became familiar with silk damasks and similar industries were set up first in Italy in the thirteenth and fourteenth centuries and in the fifteenth century in France, centred on Lyons. It was, however, in the more northerly weaving centres that linen damasks became established, borrowing the looms and techniques from the silk industry and using the high lustre linen yarn that the weavers were skilled and experienced in handling. Linen damask is completely reversible and achieves its effect by reflecting light off two-directional surfaces; white on white, the silvery patterns of the weft gleam on the snowy background of the warp (or *vice versa* on the reverse side). The first centre for linen damasks was Courtrai in Flanders and from there the technique spread to other places where linen was woven. In Holland fine damasks were made in Haarlem, while German damasks were made mainly in Saxony and Silesia. Early in the eighteenth century damask weaving was introduced to Scotland, initially in Edinburgh and then in Dunfermline. At about the same time it reached Ireland but not until 1765 did William Coulson of Lisburn develop the industry and later still another important damask factory was established in Belfast by Michael Andrews.

Experiments with looms during the eighteenth century resulted in the invention by Joseph Marie Jacquard, a French silk weaver from Lyons, of the loom which is named after him. Jacquard was summoned to Paris in 1801 and his loom

was completed there in 1804. Patterns woven on a Jacquard loom have a much cleaner line round the edges while drawloom patterns can be recognised by their step effect. The Jacquards for linens use the same five-shaft ground weave as the drawloom, although very fine woven damask (that is not less than 165 threads warp and weft per inch) are woven with an eight-shaft ground weave, called double damask.

The Jacquard mechanism was more readily accepted by the industrialised Lancashire cotton factories than by the Lyons silk weavers, who felt threatened by the consequent economies of labour. However, by the 1820s the mechanism was not only used for silk but was added to the looms of some Scottish and Irish linen damask weavers.

24

An eighteenth-century German linen damask woven with a white warp and a blue weft. It depicts scenes from the story of Jacob (Genesis, chapter 29). This is woven in what is termed a 'comber' repeat, that is, the pattern repeat faces the same direction across the width of the cloth, and was the more usual technique used by German weavers.

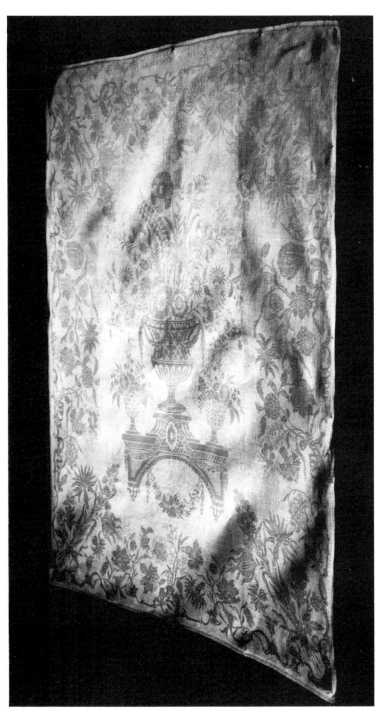

A damask napkin inscribed 'COULSON MANUFACT LISBURN IRELAND'. Probably dating from around 1800 it shows the flowing, rather formal design of the late eighteenth century.

A damask napkin handloom weaver from Waringstown, Northern Ireland. The Jacquard loom is shedded by a perforated card system, one card to each pick (row) of the pattern; a series of needles, springs and hooks lifts the individual heddles as required. The draw boy now dispensed with, the weaver is in complete control of the loom, the change from card to card achieved by depressing a single treadle. Jacquard weaving by hand continued side by side with power loom weaving, since the fabric produced by the former was considered to be better in stability, texture and strength.

ABOVE: *Bleaching linen in Northern Ireland early in the twentieth century. The laying out of cloth on bleach greens continued until the 1930s, vanishing completely after the Second World War.*

BELOW: *With the introduction of water-driven machinery, the Irish developed their own method of finishing. William Hincks's engraving of 1783 shows the procedure inside the mill. The cloth was boiled in large vats and between each boil was thoroughly washed while being pounded by the (water-driven) heavy wooden mallets, a direct adaptation of the 'tuck' mill or 'fulling' mill used in the woollen industry. The final thorough cleansing was through rub boards. The cloth, being formed into a continuous rope and after being well soaped, was passed between two grooved hard surfaces moving across each other to create a rubbing action. It was harsh treatment suitable only for the heavier types of linen.*

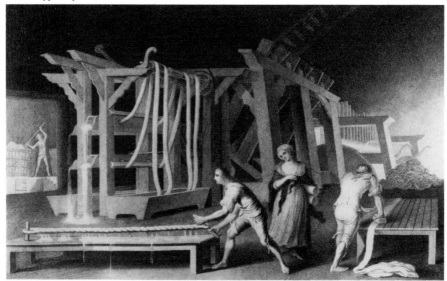

The beetling engine produced a smooth hard sheen on the surface of the cloth. The linen was wound on to a cylindrical cloth beam which slowly rotated as the heavy beechwood beetles were released, fell, bounced on the cloth and were raised again by the succeeding 'wiper' or 'cam' attached in two complete spirals on the 'wiper' beam above. Beetling lasted from a few hours to two weeks according to the type of cloth and the finish required. It made a thunderous noise, causing many of the workers to become deaf.

FINISHING LINEN CLOTH

An important part of making linen is the bleaching process. The decolourisation of the natural impurities in linen through oxidisation from the atmosphere and energy from the sun has proved to be the safest way to bleach linen for thousands of years. Pliny tells us that white linen was more highly esteemed than coloured cloth and that Egyptian linen was admired for its whiteness.

There were basically three processes in bleaching, which, after an initial steeping, were repeated over and over again until the desired degree of whiteness was obtained. These were: *bucking,* boiling the cloth in lye (alkalised water); *grassing* or *crofting,* laying the cloth out on fields to expose it to air, sun and dew; *souring,* soaking the cloth in a weak solution of acid to act as a neutraliser. In some methods souring was a final process only. Alkalis that were used included wood ash

from various sources, fern and seaweed ash (kelp), though this last was inclined to leave a yellowish tinge. Lime was used extensively for bleaching but there were laws against it because if the process was not thoroughly understood it could be harmful to the cloth. Soap and soda were also used for the lye. Souring was done with buttermilk, sour milk, water fermented with bran or rye meal, vitriol and weak sulphuric acid. These could also damage cloth if not completely rinsed away.

The cloth, when laid out on the bleach green, was kept damp all the time so that the lye could take effect but not damage the cloth. The length of time it was exposed varied from two to fourteen days. Since the cloth was vulnerable to thieving, guards were mounted, and in some places small watch towers were built.

29

Haarlem in Holland was famous for bleaching since the water in the local lake was filtered by the surrounding sand dunes, making it very pure and soft. Until the end of the eighteenth century it was a flourishing business, the whole operation taking from six to eight months. Linens from many places, including Scotland, were sent to Haarlem to be bleached.

Yarn, when it has been spun, is usually given a preliminary boil to soften and lighten the colour; in the eighteenth century there were special bleachfields for yarn. Households producing linen for their own family would do the bleaching themselves or take it to the nearest bleach works. Independent weavers who wove their cloth in the 'brown state' would sell their bolts of cloth either direct to bleachers or to merchants who were responsible for the bleaching and marketing.

At the end of the eighteenth century chloride of lime (bleaching powder) was discovered. This quickened the bleaching process but, if not properly handled, it, too, could cause linen to turn yellow after some time. Modern methods employ sodium carbonate to remove impurities and hydrogen peroxide to bleach, all processes taking place inside the factory.

The final finish on linen was produced by pounding, or *beetling,* it to close the surface of the cloth and bring out the natural lustre of the fibre, making it smooth and glossy. The Romans used glass rubbers (and also screw-down presses), while in the Swiss linen industry of St Gallen (flourishing between the fifteenth and seventeenth centuries), large marble balls were rolled over the linen to smooth it. Today beetling has been superseded by calendering machines for finishing linen. These are large metal revolving cylinders, sometimes heated, and with adjustable pressures.

Dyeing linen takes place after bleaching and has always been difficult; the hardness of the fibre resists penetration. Only the strongest of the natural water-soluble dyes, such as madder (*Rubia tinctoria*) for shades of pink and red, weld *(Reseda luteola)* for yellows and logwood (*Haematoxylon campechianum)* for blacks and greys, are satisfactory for linen, but with frequent washing they gradually fade. Vat dyes are the best for linen and by far the most frequently used natural dye is indigo (*Indigofera tinctoria),* giving various shades of blue. Before synthetic dyes were developed printed linen was mainly in striking designs of blue and white, also using indigo. Substantive natural dyes which are used for linen with some success are walnut (*Juglans nigra*) and butternut, also known as white walnut (*Juglans cinerea*), found in North America. Butternut and indigo were used in the earlier days of the American settlers. With improved synthetic dyes, linens can now be dyed with a wide range of fast colours.

Because linen can withstand frequent washing, it was useful for the many items that soil easily. In seventeenth-century England washing days were not, in many households, a regular weekly chore, but an event that took place once a month or even once a quarter. It was therefore necessary to have sufficient linens to last from one wash day to the next, so that having infrequent laundry days became a status symbol. The laundering of linen was similar to the finishing processes after the cloth has come off the loom, but not so vigorous. The hard sheen was produced by the use of starch, which also protected the cloth from dirt penetration. Frills and pleats would disappear with each washing and were restored by using goffering irons and crimping rollers while the linen was still damp. Before irons were introduced, mangles were used to smooth the linen.

Linen was stored in chests to keep it clean and free from dust. An airing cupboard with hot pipes running through was not suitable since the warmth tended to cause yellowing of the fabric.

Few households now use linen as an everyday commodity and old linens are becoming collector's items. Pieces should be kept clean and unless very frail or stained (when a professional conservator's advice should be sought) no harm will be done if the linen is given a careful wash in warm water with a non-coloured, bleach-free detergent such as Synperonic N; thorough rinsing is essential. The cloth should be smoothed out and dried flat and, when still quite damp, pressed with

30

The laundry mangle consisted of two smooth rollers turned by a handle, but an earlier version was the hand mangle consisting of a board and roller. This example comes from Denmark, where the upper sides of the boards were often decorated with carvings, floral designs, scrolls, geometric patterns and the date. Similarly decorated boards were also produced in Norway and Scotland. The linen was wrapped round the roller and worked to and fro by means of the boards, the laundress pressing down as hard as possible.

a fairly hot iron, or a steam iron may be used. Linen damasks should not be ironed but dried flat on a marble top or Perspex sheet. This brings up the sheen on the right side (leaving the back rather rough), showing the pattern to best effect. When dry, the linen should not be folded, but flat pieces should be covered with acid-free (this should be stated on the pack) tissue paper and rolled round plastic tubes.

LINEN TODAY

During the early 1950s most branches of the linen industry suffered further decline, mainly due to the widespread use of man-made and synthetic fibres for making easy-care fabrics. Although much admired, the 'linen look', even after treatment with crease-resistants, still required ironing.

The canvas industry using dry-spun yarn has been less affected. Linen has the advantages of being stronger and having a longer life than cotton; it lacks the problems of condensation, a common complaint with synthetics; and it is, by the nature of the fibre, resistant to tearing. Since the fibre swells when wet there is a degree of natural waterproofing, and chemical treatment for water, rot and flame proofing can nowadays be achieved easily and cheaply. Mail bags, awnings, barge and lorry covers, horse rugs, camping equipment, water bags and luggage are some of the uses for canvas today.

In the factories modern mechanical technology, with improved dust extraction and enclosed machinery, has not only made conditions safer and healthier for the workers but has increased the efficiency of the machines. Rapier looms, which substitute flexible rods for shuttles, have permitted a fourfold increase in productivity, although spinning by the 'open-end' system, the most modern method used in the cotton industry, is not suitable for flax because of the dirt and wax content as well as the length of the fibres.

Much research continues into crease-resistant processes, which are at their most successful when flax is blended with other fibres such as polyester and viscose as well as wool and cotton. Abrasion is another problem which is reduced by

31

blending. However, furnishings, particularly where drapability is an important asset, and wall coverings, where creasing is not a concern and where surface texture and aesthetic considerations are of primary importance, make use of linen.

Linen sheets are now a luxury few can afford but other household items such as drying-up cloths, which polish glass so easily and efficiently, and decorative table mats continue to find a market, as also do fine lawn handkerchiefs.

Although this oldest of textile fibres now accounts for only 2 per cent of the world's textile production, clothing designers have turned again to using fabrics made from natural fibres and linen is being chosen for its comfort and good looks, particularly in hot climates where it has regained prestige value.

FURTHER READING

Clark, Wallace. *Linen on the Green*. The Universities Press (Belfast) Ltd, 1982.
Crowfoot, Grace M. *Methods of Handspinning in Egypt and the Sudan*. Ruth Bean, reprint 1974.
Durey, Alastair. *The Scottish Linen Industry in the Eighteenth Century*. John Donald (Publishers Ltd), 1979.
Evans, Nesta. *The East Anglian Linen Industry and Local Economy 1500-1850*. Gower Publishing Company Ltd, for Pasold Research Fund, 1985.
Geijer, Agnes. *A History of Textile Art*. Pasold Research Fund, 1979. (Chapter 9, pages 171-8, 'Linen damask and other table linen'.)
Ulster Folk and Transport Museum. *Illustrations of the Irish Linen Industry in 1783 by William Hincks*. HMSO (Northern Ireland), 1977.
Worst, Edward. *How to Weave Linens*. The Bruce Publishing Company, 1926.

PLACES TO VISIT

Many museums in the British Isles offer one aspect or another of linen, though few cover the subject comprehensively, there being greater interest in the wool industry. In mainland Europe small town museums are far more likely to show items relating to flax production, spinning wheels, scutching tools etc. Folk and costume museums, both in the British Isles and elsewhere, are always worth a visit, as many exhibits are likely to be made from linen. The one museum that is totally devoted to flax technology is at Courtrai in Belgium. Intending visitors are advised to find out the times of opening before making a special visit to a museum.

UNITED KINGDOM

The British Museum, Great Russell Street, London WC1B 3DG. Telephone: 01-636 1555 or 1558. (Egyptian Galleries.)
Dunfermline District Museum, Viewfield, Dunfermline, Fife KY12 7HY. Telephone: Dunfermline (0383) 721814. (Local damask industry.)
Lisburn Museum, The Assembly Rooms, 14 Market Square, Lisburn, County Antrim, Northern Ireland. Telephone: Lisburn (084 62) 72624. (Local linen industry.)
Science Museum, Exhibition Road, South Kensington, London SW7 2DD. Telephone: 01-589 3456. (Textile technology.)
Ulster Museum, Botanic Gardens, Belfast, Northern Ireland BT9 5AB. Telephone: Belfast (0232) 668251-5. (Linen technology.)
Victoria and Albert Museum, Cromwell Road, South Kensington, London SW7 2RL. Telephone: 01-589 6371. (Textiles.)

OTHER COUNTRIES

National Vlasmuseum, Etienne Sabbelaan 4, Courtrai (Kortrijk), Belgium.
Twente and Gelder Textile Museum, Espoortstraat 182, Enschede, Overijssel, Holland.